OUR UNIVERSE

A Primer About Matter, Energy, and How We Know What We Know

Harry Gilbert

To Barbara
sine qua non

Table of Contents

Preface

This book has grown from notes I have prepared over the years for my eighth grade science class. It now provides a reference for my students before and after various labs, activities, and conversations during class. So, the language of this book is often as if I am addressing my students, and this book does not include any of the other materials we use to illustrate, explain, or extend the basic concepts noted here. As a result, some sections may seem incomplete; however, basic concepts are covered.

Acknowledgements

Thank you to Diane Russell Harbison, Richard C. Henry, and Bruce A. Mason for reviewing original versions of sections III, II and V, and IV, respectively, which helped me correct errors and sloppiness in my generalizations among other faults (any remaining errors are mine). I also thank Deborah K. Watson for obtaining and coordinating the reviews by Professors Henry and Mason. Thank you to Allison Gilbert for her comments that improved part I, and thank you to Sue Oldham for her overall review.

I. Before We Begin – Some Basic Science Skills and Information

Before we study our universe, we need to learn about or review some of the tools of science that help us make observations and record information about our world and universe. These tools include measuring, organizing information, presenting information, and using math. We have already worked with various pieces of scientific equipment and we will continue to do so this semester. Near the end of the semester, we will review formal lab reports. Although we will study these throughout the semester, they are all included in this section I.

First, however, what does the term "scientific method" mean?

A. scientific method

To begin, we make an assumption that is one of the basic ideas of modern science: the universe is regular and predictable. This is the basis of scientific method and its idea of repeatable testing.

Although scientific method is often described by the parts of a formal lab report, which we will discuss later, I use three words to describe a version of scientific method: observe, question, and test.

We observe something, and it raises questions such as – What is that? or What caused that? or How did that happen? or ….

We also draw inferences from our observations and make hypotheses, or possible answers to our questions. Based on our hypotheses, we make predictions about events that we can test. We then test (such as through experiments) to see if our predictions and hypotheses are supported by the results of the test. Records of the testing procedure and results are made so the test can be repeated. This testing of our predictions and hypotheses to see if they are supported or refuted distinguishes science from other human disciplines, such as philosophy and theology.

Such testing can lead to new observations and questions, and so this scientific process continues: observe, question, test, observe….

Now, on to measuring, organizing and presenting information, using math in science, and writing lab reports.

B. measuring

1. measuring equipment and units

We will work with meter sticks, mass balances, stopwatches, and thermometers with which to measure distance, mass, time, and temperature. Each of these pieces of equipment has a scale that gives us our measurement reference and unit. Some common units for distance, mass, time, liquid volume, and temperature are meter (m), gram (g), second (s), liter (L), and °Celsius (°C), respectively.

2. significant figures

Counting a number of objects typically gives a definite whole number result, such as 14 students in a class. On the other hand, measuring some property of an object, such as length or volume or temperature, involves some degree of uncertainty as we compare the property to a measurement scale. The scale that we use limits the precision* with which we can report the measurement that we make. This is where significant figures come in. [*The accuracy, as opposed to the precision, with which we measure something depends on the quality of the scale and how we use it. For example, a broken meter stick, or one that has a worn end, or not carefully lining up or reading the scale with what is being measured affect the accuracy of the measurement.]

Significant figures are those digits in a measurement that are meaningful, based on the measurement scale that is used. These digits are the ones we know for certain (based on the scale markings of the measurement tool we are using) plus one additional digit, which is estimated. For example, a meter stick typically has 1,000 marks along its length, each mark defining 1 millimeter (1 mm). So, we can measure a length with certainty to the nearest millimeter (three decimal places to the right of the decimal point). We can, however, also estimate between two adjacent millimeter marks if the length we are measuring falls between them; therefore, we can add one more digit to our measurement (the fourth place to the right of the decimal point for our millimeter example).

At the end of the year we will consider significant figures when making a measurement relative to some defined scale, such as the markings on a meter stick, graduated cylinder, or thermometer. We will also look at rules that help us determine how many significant figures are in a given measurement and rules that apply when a measurement needs to be rounded off. There are still more rules to apply when we have to add or subtract measurements and different rules when you multiply or divide measurements. I will give you a separate handout about these rules when we get to them.

2

C. organizing and presenting information

Once you have measured, or made other types of observations, you might need to organize the information to make sense of it. Six organizational tools to help you do this include concept maps, compare/contrast tables, Venn diagrams, flowcharts, cycle diagrams, and data tables. You are probably already familiar with at least some of these.

You can also organize by making graphical presentations with bar graphs, line graphs, and circle graphs. You have also used these before, but we will review them along with concept maps, compare/contrast tables, Venn diagrams, flowcharts, cycle diagrams, and data tables.

D. using math in science

You will use math throughout your science classes –estimating, doing basic arithmetic (adding, subtracting, dividing, and multiplying), solving equations, and more that you will learn in high school. For now, we will learn about numerical prefixes, powers of 10, math with units, conversions, and word problems

1. numerical prefixes

Meter, gram, and liter are examples of units of measurement. Each relates to what is being measured. Meter relates to distance, gram relates to mass, and liter relates to volume, for example.

A numerical prefix may be used with a basic unit of measurement to specify a multiple or a fraction of that unit. For example, kilogram means one thousand grams and centimeter means one-hundredth of a meter.

In the table of numerical prefixes that you have completed in class, you will see the prefix, such as tera- and its symbol (T), its meaning (one trillion) and the power of 10 (10^{12}) representing this meaning. For this example using tera-, a terameter = a trillion meters = 1,000,000,000,000 meters = 10^{12} meters. So, to change from a numerical prefix to only the base or root unit, just delete the prefix and substitute the power of 10 that corresponds to the deleted prefix: for example, 6 terameters = 6 · 10^{12} meters because the prefix "tera" means 10^{12}.

2. powers of 10

a. in general

A power of 10 has the form 10^x. The 10 is the base and x is the exponent (also called the power, such as 10^2 being "ten to the power 2" or "ten to the second power"). In our work, x will be an integer (…-2, -1, 0, 1, 2, …).

If x is 0, $10^0 = 1$ (any base to the 0 power equals 1).

If x is a positive integer, 10^x represents the number 1 followed by x zeros. For example, $10^3 = 1$ followed by 3 zeros = 1,000.

If x is a negative integer, 10^x represents the fraction $1/(10^{|x|})$. In decimal form this fraction is written beginning with a decimal point, followed by $|x| - 1$ zeros, followed by the number 1. For example, $10^{-5} = 1/(10^5) = 0.00001$.

b. powers of ten multiplied by other factors

Powers of 10 can have factors associated with them. That is, the number you are working with might include a non-power-of-ten number multiplied by a power of ten (for example, 29 x 10^3). Scientific notation puts these factors in a particular form (see subsection e below).

c. multiplying and dividing

To multiply powers of 10, add the exponents. For example, 10^2 x $10^6 = 10^{2+6} = 10^8$. So, the general rule for multiplying powers of 10 is 10^a x $10^b = 10^{a+b}$.

To divide powers of 10, subtract the denominator's exponent from the numerator's exponent. For example, $10^7/10^3 = 10^{7-3} = 10^4$. So, the general rule for dividing powers of 10 is $10^a / 10^b = 10^{a-b}$.

If there are other factors, they are multiplied or divided separately from the powers of 10. For example, 2 x 10^4 x 3 x $10^5 = (2$ x $3)$ x $(10^4$ x $10^5) = 6$ x 10^9 .

d. adding and subtracting

To add or subtract numbers with powers of 10, the powers of 10 must be the same for all the numbers to be added or subtracted. So, you might have to move the decimal point on a number that is multiplied by a power of 10 and change the power of 10 exponent. Once the powers are the same, add or subtract the factors and multiply that sum or difference by the common power of 10. For example, (22 x 10^5) + (3 x 10^6) = (2.2 x 10^6) + (3 x 10^6) = 5.2 x 10^6.

e. scientific notation

Scientific notation is simply a specific power-of-ten format, namely,

M x 10^x, where $1 \le |M| < 10$.

For example, 34 x 10^4 is in a power-of-10 format, but it is not in scientific notation format. This value in scientific notation format is 3.4 x 10^5.

3. math with units

To add or subtract measurements (values having numbers and units), the units must be the same before the numbers can be added or subtracted (to do this, you might have to do a conversion as explained below). Examples:

$$2 \text{ seconds} + 3 \text{ seconds} = 5 \text{ seconds}$$

$$2 \text{ meters} + 3 \text{ seconds} = 2 \text{ meters} + 3 \text{ seconds (can't be reduced)}$$

To multiply or divide measurements, the units are multiplied or divided and so they do not have to be the same root unit (but prefixes of the same root need to be the same). If the units are the same and they are multiplied, you get unit^2 (or cubed, etc.). If the units are the same and they are divided, the units cancel. Examples:

$$2 \text{ meters x } 3 \text{ seconds} = 6 \text{ meter} \cdot \text{seconds}$$

$$2 \text{ seconds x } 3 \text{ seconds} = 6 \text{ seconds}^2$$

$$8 \text{ seconds } / 4 \text{ seconds} = 2$$

4. conversions

Sometimes you need to change a measurement from one unit to an equivalent value in another unit. For example, maybe something is stated in meters but you want to know what the measurement is in centimeters. Or maybe you want to change from hours to seconds. To do this, you multiply the beginning measurement by the appropriate value of 1 to get an answer in the desired units.

Here is an example of a conversion:

Let's convert 2 meters to centimeters. To do this, we ask: What do we have to multiply the measurement "2 meters" by to convert this measurement to the equivalent number of centimeters? Because the question refers to meters and centimeters, and since 1 meter = 100 centimeters, let's multiply by 100 centimeters/1 meter, which ratio equals 1. This gives us: 2 meters x 100 centimeters / 1 meter = 200 centimeters

Because the same unit, meter(s), is in both the numerator and the denominator, this unit cancels (see section 3 above). Because the meter(s) units cancel, this leaves centimeters as the unit in the answer.

The value of 1 that we choose (100 centimeters/1 meter in the example above) is called the conversion factor. The conversion factor has a value of 1 because the value of the numerator equals the value of the denominator (that is, 100 centimeters = 1 meter in our example). So, multiplying the original measurement by 1 does not change its value, just the form we express it in.

Here is a procedure for converting a measurement from one unit to another unit:

1) Near the left margin of your paper, write the measurement you want to change.

2) Make the conversion factor by doing the following four steps a-d:

a) To the right of the original measurement you just wrote near the left margin, write a multiplication symbol followed by a horizontal division symbol (x --------------).

b) Identify the name of the unit to be changed. Write that unit name above or below the horizontal division line <u>opposite</u> to where that unit is in the original form of the measurement that you wrote in step 1. Put this unit name toward the right end of the division line.

c) On the other side of the division line, write the name of the unit you are changing to. Put this unit name toward the right end of the division line.

d) To the left of the two unit names, fill in the correct numbers that make this conversion factor have a value of 1 (that is, the number and unit in the numerator equals the number and unit in the denominator).

3) Put an equal sign (=) to the right of the conversion factor and do the math to complete the conversion.

5. word problems

When you're given various pieces of information and asked to calculate something from the given information, do the following:

1) write down what you are asked to calculate ("find," "determine," etc.), then

2) write down the information you are given or know is needed to calculate what you wrote down in step 1. State the word(s) for the information and the value (including units), then

3) write the general equation that the information calls for, then

4) plug the given information (numbers and units) into the general equation to set up the specific equation to solve, and then

5) use your math knowledge to solve the specific equation for what the problem says to find (see step 1 above).

E. writing lab reports

As we do labs and activities throughout the semester, note how some of the information is laid out for you on the instruction sheets you are given. These instruction sheets follow many of the sections that you will put into formal lab reports that you write in high school. At the end of this semester we will review this format.

Lab reports you write in high school should, of course, follow the format your teacher tells you to follow; but this will likely have the content listed on the handout that you will be given near the end of the year. You might not always do all the parts, but these are the parts of a complete report when you have to write one. In addition to what's on the handout, remember the following:

1) details, details, details! (understand the details of the lab, observe details, report details)

2) read and understand instructions <u>before</u> you go into the lab

3) record what <u>actually</u> happens - fully and accurately

4) record using words and pictures as appropriate

5) talk to your teacher before you turn in a lab report.

II. We Had a Beginning – From Then Until Now (a little cosmology)

In high school you will study biology, chemistry, and physics. The topics in each of these subjects will mostly be about life, matter, and energy here on Earth.

But what about the rest of the universe, and where did all this stuff come from, and how do we know?

A. the Big Bang

1. background

Until a few hundred years ago, humans thought we were the center of the universe. We thought all the planets and stars revolved around Earth. This geocentric (earth-centered) idea began to change with Copernicus's 1543 publication of a book that argued that the sun, not Earth, was the center of at least our solar system. It took decades for this heliocentric (sun-centered) idea to take hold.

In the early 1600s support for the heliocentric idea came from Kepler and Galileo. Kepler figured out that planets have elliptical orbits around their stars. This contradicted the geocentric belief that objects in space orbited in perfect circles. Galileo used a telescope to study the night sky, and one of his observations was of Jupiter's moons. This showed that objects revolve around things other than Earth.

Over the past 400 years more and more evidence has established that we live in a heliocentric solar system which does not appear to be at the center of anything. We are far out in one spiral of a galaxy called the Milky Way, which is only one of billions and billions of galaxies in our universe.

Our knowledge about the existence of other galaxies is fairly recent. Less than 100 years ago (in the mid-1920s) an astronomer named Edwin Hubble discovered that objects we thought were in our galaxy are actually far beyond it. The objects that Hubble observed were some of the billions and billions of other galaxies in our universe. Take a look at, for example, the Hubble Deep Field photographs on the internet to see a tiny fraction of these galaxies.

2. expansion of the universe

A few years after Hubble discovered that there are galaxies beyond our Milky Way galaxy, he measured galactic distances and put them with velocity information to discover something that has had an even more profound effect on our understanding of our universe: distant galaxies are moving away from us, so our universe is expanding. More specifically,

Hubble discovered that these galaxies are moving away from us according to an equation now known as Hubble's law: $v = H_0 d$, where v is the speed a galaxy is moving away from us, H_0 is a number known as the Hubble constant, and d is the distance to the galaxy. This implies that our universe was smaller in the past, back to a beginning point and time. This discovery by Hubble was initial observational evidence supporting what is now called the Big Bang theory.

3. cosmic microwave background radiation

Arno Penzias and Robert Wilson discovered the cosmic microwave background radiation in 1964. This radiation was created by the heat produced at the beginning of our universe. We can still detect this radiation today.

One second after the beginning event, the temperature of the universe was about 10^{10} K (ten billion kelvins). This created very high energy radiation (think of this radiation as waves of energy). As the universe has expanded, however, the universe has cooled and the energy of the heat radiation has decreased, so the temperature has gone down. One prediction of the Big Bang theory is what this temperature should be today. Penzias and Wilson discovered this radiation and it was indeed at this lower temperature, thereby experimentally confirming this theoretical prediction of the Big Bang theory.

B. evolution of our universe

1. $E = mc^2$

So, what happened at that beginning, at the instant of the bang of the Big Bang?

Well, first, it wasn't an explosion; it was just a sudden beginning to an expansion of space that we now know as our universe. But other than this sudden expansion, we don't know what happened at the specific beginning moment. Our current scientific knowledge and math do not enable us to explain that beginning instant. However, we can get pretty close. Energy and matter that we observe today can be traced back to within a second of the Big Bang.

At the beginning itself, possibly something caused a massive release of energy, part of which changed to matter. This convertibility of energy to matter (and matter to energy) is expressed in Einstein's simple but profound equation, $E = mc^2$ (energy equals mass times the square of the speed of light). This relationship came out of Einstein's theory of special relativity in 1905.

Whatever happened at the beginning instant, space, time, energy, and matter have evolved to what we now observe today: an expanding and accelerating universe about 13.8

billion years old that includes about 5% ordinary matter, 27% dark matter, and 68% dark energy. After we look at a little of this evolution of our universe, we will look more closely at ordinary matter and energy around us before we return to the universe at large.

2. elementary particles

The elementary particles that came out of the Big Bang within its first second are the quarks and the leptons. We presently know of three families of these particles (and their anti-particles).

class of particle	family 1	family 2	family 3
quarks	up down	charm strange	top bottom
leptons	electron electron neutrino	muon muon neutrino	tau tau neutrino

Only the members of the first family (the up and down quarks and the electron and electron neutrino) make up ordinary matter of the type that we encounter every day. Specifically, two up quarks and one down quark make one proton; one up quark and two down quarks make one neutron. Electrons are one type of elementary particle in the group we call leptons.

3. fundamental forces

As energy and matter separated during the first second of the Big Bang, four fundamental forces that interact with matter also separated: gravity, weak nuclear force, electromagnetic force, and strong nuclear force.

Gravity is an attractive force between objects which have mass. At a comparable scale it is the weakest of the four forces. According to theory, this force gets transferred by gravitons, one type of a class of particles called bosons.

Next weakest, but still substantially stronger than gravity, is the weak nuclear force (or, simply, weak force). The weak nuclear force acts within nucleons (protons and neutrons) whereby a neutron can change to a proton and a proton can change to neutron (other particles also come out of these transformations). This force includes bosons called W and Z particles.

A few powers of 10 stronger than the weak force and about 100 times weaker than the strong force, the electromagnetic force causes the electrical attraction between a positively charged particle and a negatively charged particle (and the repulsion between like charged particles). This force includes the force of magnetism; and it is fundamental to electromagnetic waves, which have interrelated electric and magnetic fields. This force includes boson particles called photons.

The strong nuclear force, or simply strong force, binds quarks to form protons and neutrons, for example, and to hold protons and neutrons together in atomic nuclei. This force includes bosons called gluons.

4. nuclei for atoms

So, we're only within a second of the Big Bang and yet we have quarks, leptons, and the four forces. Quarks, under the influence of the strong nuclear force, then combined to make protons and neutrons. Protons and neutrons combined to make nuclei for what would eventually become atoms that make up the matter around us today. These nuclei have formed through nucleosynthesis processes.

The term "nucleosynthesis" refers to making atomic nuclei that are heavier than a single proton. A single proton defines the simplest element, which we call hydrogen. However, nucleosynthesis includes making a variety of hydrogen, called deuterium, which has both a proton and a neutron. Helium through uranium are other elements whose nuclei have come from naturally occurring nucleosynthesis.

Most of the deuterium and helium nuclei were made within a matter of minutes of the Big Bang. Because the word "primordial" means "first in time," this creation of atomic nuclei is called primordial nucleosynthesis.

Heavier nuclei formed, and are still being formed, in processes that occur in stars (the word "stellar" refers to stars so this is called stellar nucleosynthesis). Uranium is the heaviest element made by these natural stellar nucleosynthesis processes.

a. primordial nucleosynthesis

The universe cooled as it expanded from the moment of the Big Bang. As this cooling occurred, the strong force joined quarks together to form protons and neutrons.

At cooler temperatures, but still within seconds of the Big Bang, the strong force bound neutrons with protons to form deuterium nuclei (each deuterium nucleus has one proton and one neutron). These nuclei in turn combined to form helium nuclei; each of which has two protons and two neutrons. In less than 15 minutes after the Big Bang, the universe had cooled further such that there was insufficient energy to keep this going so these nuclei-creating reactions stopped. It was, however, still hot enough to keep electrons from binding with these nuclei; it would take 380,000 years before the universe was cool enough for electrons to stay bound to nuclei and form the first atoms.

b. stellar nucleosynthesis

Whereas primordial nucleosynthesis happened as the beginning universe cooled, stellar nucleosynthesis occurred (and continues to occur) in stars. The first stars did not exist until about 300 million years after the Big Bang. New stars continue to form today.

To make a star, matter (primarily hydrogen gas) in space comes together under the force of gravity. As more and more matter gathers, the density and temperature increase until it is hot enough for nuclear fusion to begin – a star is born.

In the nuclear fusion that occurs in the core of the star, hydrogen nuclei fuse to make helium nuclei. These helium nuclei combine to make nuclei of even heavier elements, which in turn make still heavier nuclei as the temperature in the core of a star gets hotter and hotter. This can continue in some stars until iron nuclei are formed. Different layers of these nuclei build up, with the lighter elements pushed outward in the star and the heaviest at the center of the core. During the end stages of the star, such as when a star like our sun becomes a red giant and then a white dwarf, these elements get hurled into space.

Elements having nuclei heavier than iron form in nova and supernova events, explosive processes that can occur with white dwarfs and massive stars. In a nova event, a white dwarf gathers matter from another star that is locked by gravity in orbit with the white dwarf. The accumulated matter includes lighter element nuclei. The amount of these elements builds up on the surface of the white dwarf until it is so dense and hot that the accumulated matter explodes out into space. Sometimes the density and temperature can be so great as to start processes by which the white dwarf itself also explodes, producing one type of supernova. Either of these events produces enough energy to fuse lighter nuclei together to create nuclei heavier than iron.

In another type of supernova, a sufficiently massive star (one several times more massive than our sun) itself directly explodes with enough energy to produce elements up to uranium.

5. particle accelerators

Particle accelerators accelerate subatomic particles (e.g., protons, neutrons or electrons) to very high speeds and then guide these high-speed particles into collisions with each other.

These collisions create a variety of other particles and energies that enable scientists to discover different types of matter and antimatter made from quarks and leptons. It is from these experiments that scientists have confirmed the families of quarks and leptons listed in the table above.

Also, it is from such experiments that scientists hope to find the answer to the question of what gives elementary particles the property we call mass. The current particle physics theory has predicted that mass arises from a field associated with a particle called the Higgs boson. Recent experiments at the Large Hadron Collider on the French-Swiss border have shown that this prediction is correct; the Higgs boson exists. Two scientists who predicted the existence of the Higgs boson won the 2013 Nobel Prize in Physics.

III. The (Almost) 5% Universe – Ordinary Matter (a little chemistry)

    A. introduction

Carbon, nitrogen, oxygen, sodium, phosphorus, sulfur, and iron are a few of the atoms that make up our bodies. Aside from primordial hydrogen, these and other elements in our bodies have literally come from the stellar nucleosynthesis processes described previously.

These atoms in our bodies are examples of the 92 different elements that have been made naturally throughout the history of our universe. These 92 elements make up the (almost) 5% ordinary matter of our universe. By "matter" I mean anything that has mass and occupies space. By "ordinary" I mean the matter that we can observe directly; this is also referred to as atomic matter because it is made of the atoms listed on the Periodic Table of the Elements. The other type of matter, dark matter, seems to have mass and to occupy space, but we have not been able to observe it directly; we can only observe its effects. We will study only ordinary matter in our chemistry unit and so this is what the word "matter" by itself will refer to in this part III.

Chemistry is about the composition, structure, and properties of matter and the chemical changes this matter undergoes. A significant part of chemistry is to understand the structures of the atoms that make up the elements and how these structures affect whether one element will chemically react with another element. As we will see, chemical change occurs through chemical reactions, which involve the electrons of atoms and the electromagnetic force. Chemical reactions are different from the nuclear reactions we previously discussed; remember that those nuclear reactions occurred in nuclei, which are at the centers of atoms, and involved the weak and strong forces, protons, and neutrons.

    B. properties of ordinary matter

Matter has mass and weight, which are different but related concepts. Mass relates to the amount, or quantity, of matter in something. Weight relates to the force exerted by gravity on a mass. Mass is independent of gravity; weight depends on gravitational acceleration at the place where the weight is being measured. So, your mass is the same on Earth and the moon, but your weight is different on each because the gravity of each is different.

Matter most commonly exists here on Earth in any of three classical states or phases: solid, liquid, or gas. A solid has a definite shape and volume. A liquid has a definite volume but not a fixed shape – it can flow so a liquid takes the shape of its container. A gas does not have a fixed shape or volume; it takes the shape and volume of its container. A gaseous system having positively charged particles and negatively charged electrons not bound to atomic nuclei is called a plasma, which can be considered a fourth state or phase of matter.

C. classifications of ordinary matter

The ordinary matter we will consider is either a pure substance or a mixture. We will study pure substances first.

1. pure substances

A pure substance is matter that has the same composition and properties throughout a given sample and from sample to sample of the substance. A bar of pure gold is a pure substance because it has only gold atoms in it – however you analyze the bar or cut it up, each portion will have the same composition and properties. Any other bar (or other object) of pure gold will also have the same composition and properties. Pure water is a pure substance because the water has only molecules of H_2O and so every portion of the volume of water will have the same composition (only H_2O molecules) and properties; any other sample of pure water will be the same.

Pure substances are either elements or compounds. An element has a single kind of atom (gold, for example). A compound includes two or more kinds of atoms chemically bonded together (H_2O, for example).

a. elements

(1) atoms – general structure, ions, and isotopes

Every atom of a particular element has an identical number of protons in the nucleus of the atom. Recall that the nucleus of an atom is centrally located in the atom. The nucleus of any atom contains positively charged protons and electrically neutral neutrons held together by the strong nuclear force (except for the basic form of hydrogen, which has one proton but no neutrons). Negatively charged electrons move around outside the nucleus; these electrons are bound to the positively charged nucleus by the electromagnetic force. Electrons orbit the nucleus in specific regions called shells or energy levels.

If the number of electrons (each with a -1 charge) orbiting the nucleus equals the number of protons (each with a +1 charge) in that nucleus, the atom is electrically neutral because the total negative charge equals the total positive charge. For example, a neutral atom of oxygen has eight protons in its nucleus (for a total positive charge of +8) and eight electrons orbiting that nucleus (for a total negative charge of -8), thus the net charge is 0 (+8-8=0). An atom of that element having a different number of electrons is still the same element because the number of protons is the same; however, the atom is no longer electrically neutral. This atom that is not electrically neutral is called an ion. An ion can have a positive charge (fewer electrons than protons), or an ion can have a negative charge (more electrons than protons). For example, an oxygen atom always has eight protons in its nucleus; however, if that oxygen atom has ten electrons orbiting the nucleus, the overall charge is -2 (+8-10= -2).

Although the atoms of a particular element have the identical number of protons in their nuclei, these atoms can have different numbers of neutrons in those nuclei. Two atoms having the same number of protons but different numbers of neutrons are called isotopes of the element. Both atoms are still the same element because the number of protons is the same in each; they are just different varieties (isotopes) of the element. For example, there are three isotopes of hydrogen, one having one proton and no neutrons, a second having one proton and one neutron, and a third having one proton and two neutrons.

(2) electron orbits, and some history

J.J. Thomson discovered electrons over 100 years ago. Because he knew typical atoms are electrically neutral, he proposed that the negatively charged electrons must be mixed with some kind of positive particles, and so he proposed what is known as the "plum pudding model" of the atom. In the plum pudding model, the electrons and positive charges are mixed together.

Over a decade later, Ernest Rutherford discovered that atoms have dense central regions. Rutherford and his assistants shot alpha particles (helium nuclei) at a thin sheet of gold and observed how the alpha particles were sometimes deflected by the atoms inside the gold sheet. From this experiment Rutherford proposed his model of the atom: a positively charged nucleus surrounded by negatively charged electrons.

Thomson's model and Rutherford's model were only steps toward our modern model of the atom. Today's model of the atom began with what is known as the Bohr model, named for Niels Bohr who combined ideas from quantum physics with the Rutherford model. Rutherford's atomic model had a central nucleus containing positively charged protons with electrons orbiting outside the nucleus. Bohr modified this model by proposing that the electrons could orbit only in specific, discrete orbits outside the nucleus. That is, the electrons cannot be at just any distance from the nucleus – they can only be at certain distances from the nucleus, with gaps in between.

As a result of Bohr's work and the subsequent development of quantum mechanics, we now understand that electrons orbit the nucleus in an atom in specific energy levels (also called shells). Each energy level is at a respective distance from the atom's nucleus, and each energy level has orbitals or subshells in which the electrons move. In our study we will only be concerned with the energy levels as a whole. As we learn about these energy levels, we will draw diagrams showing electrons as if they are points on circles outside the nucleus where each circle represents one of the energy levels in which electrons orbit the nucleus. Reality is more complicated than this, but this is the place to begin; you will learn more in high school, such as when you learn about the Pauli Exclusion Principle and the Heisenberg Uncertainty Principle.

(3) The Periodic Table of the Elements

All the elements that we know about are listed in the Periodic Table of the Elements. The Periodic Table lists the elements in a sequence according to the number of protons each element has in its atoms. For example, the element hydrogen (H) is the first entry because every hydrogen atom has only one proton in its nucleus. Helium (He) is next because every helium atom has two, and only two, protons. Every lithium atom (Li) has three protons, and so on.

The Periodic Table is arranged in numbered rows and numbered columns. At each row, column intersection there is a square containing information about a particular element. For example, where row 2 and column 1 intersect, we find the square for lithium (Li). So, what can we learn about this element?

(a) period and group

Each row number is the period for the elements in that row. The period of an element tells us the number of energy levels that contain electrons in a neutral atom of that element.

Each column number is the group for the elements in that column. Using the group number, we can determine how many electrons are in the outermost energy level of an electrically neutral atom. These outermost electrons are called valence electrons. For groups 1 and 2, the group number itself tells us the number of valence electrons. If the group number is 13 or greater, subtract 10 from the group number to determine the number of valence electrons. For example, the element chlorine (Cl) is in group 17; subtracting 10 from 17 gives us 7, so a neutral chlorine atom has 7 valence electrons. Groups 3-12 are a bit trickier and will be left for high school.

So, for our example of lithium, it is in the second period and first group, which means one electrically neutral atom of lithium has two energy levels with its outer level containing one electron.

(b) symbol and name

Each square of the Periodic Table has the chemical symbol for the element. For example, "Li" is the chemical symbol for lithium. The first letter of the symbol is always capitalized; the second letter (if there is one) is not.

Sometimes the square on the Periodic Table also includes the name of the element – the name "lithium" in our example. The element name is a common noun and so it begins with a lower case letter (unless beginning a sentence).

(c) atomic number

Each square of the Periodic Table lists the atomic number of the element. The atomic number corresponds to the number of protons in every atom of that element. The atomic numbers are the sequence numbers of the Periodic Table that increase by one from square to square when you read the table from left to right beginning with the first row (period). This is the number that begins at 1 for hydrogen (H) and counts upward from there (2 for helium (He), 3 for lithium (Li), etc.). The Periodic Table is organized according to the number of protons from element to element.

The number of protons an atom has determines which element we have. Since the number of protons is the same as the atomic number, we can find the element in the Periodic Table if we know the number of protons or atomic number. Alternatively, if we know the element, we can look up the atomic number in the Periodic Table. Knowing the atomic number, we know the number of protons in each of the atoms of that element. All atoms of a given element have the same number of protons and thus the same atomic number.

(d) mass number (atomic mass)

The mass number, or atomic mass, tells us the total number of nucleons (protons and neutrons) in the nucleus of an atom:

mass number = number of protons + number of neutrons

Different isotopes of an element have different mass numbers. Remember, each atom of a particular element always has the same number of protons as every other atom of that element; however, there may be different numbers of neutrons and thus different mass numbers.

The mass number is a whole number since there are only whole numbers of protons and neutrons in an atomic nucleus.

(e) charge

Because the charges of a proton and an electron are equal but opposite (+1 for a proton and -1 for an electron), the charge of an atom is the difference between the number of protons and the number of electrons in the atom:

$$charge = number\ of\ protons - number\ of\ electrons$$

So...

1) If there are equal numbers of protons and electrons, the atom has a total charge of 0 (it is electrically neutral).

2) If there are more protons than electrons, the atom has a net positive charge.

3) If there are more electrons than protons, the atom has a net negative charge.

The charge of any given atom is not directly shown on a typical Periodic Table; however, if the atom is electrically neutral, then the atomic number of the element also tells us the number of electrons because the number of electrons equals the number of protons in an electrically neutral atom.

(4) atom drawings

Once we know the above information, we can draw a diagram of an atom for a given element. Here is a procedure for drawing the type of atom diagrams we will use:

1) Use the atomic number to determine the number of protons and write that number and "p" in the nucleus.

2) Use the appropriate mass number I give you and subtract the number of protons to determine the number of neutrons. Write that number and "n" in the nucleus.

3) Use the period to determine the number of energy levels and draw that number of circles outside the nucleus to represent the energy levels in which the electrons orbit the nucleus.

4) Use the given information about charge for that atom to determine the number of electrons and place the appropriate number of dots (electrons) on the proper circle (energy level). If the atom is electrically neutral, use the group number (specifically, its ones digit) to check the number of electrons you placed in the outermost energy level. To place the electrons, begin with the innermost circle, which represents the lowest energy level, or shell. For the elements we will

consider, two electrons fill the lowest energy level and eight electrons fill each of the next two energy levels.

b. compounds

(1) definition

The next class of matter, compounds, is another type of pure substance. To form a compound, two or more elements bond together in definite proportions. The bonding occurs through a chemical reaction. For example, two hydrogen atoms will chemically react with one oxygen atom to form the compound H_2O, which is composed of two different elements (hydrogen and oxygen) combined in a definite proportion (always 2 hydrogen atoms to 1 oxygen atom).

A compound can be separated into two or more simpler substances by a chemical change. For example, water (H_2O) can be decomposed into its separate elements, hydrogen and oxygen, by chemical processes (e.g., electrolysis). Compounds cannot be separated into simpler parts by physical processes (e.g., evaporation).

(2) chemical bonds, ionic and covalent

During a chemical reaction that produces a compound, chemical bonds form between the atoms that make up the compound. We will consider two types of chemical bonds: ionic bonds and covalent bonds. An ionic bond forms by the transfer of one or more electrons between the atoms, thereby forming positive and negative ions which are bound together because of their opposite charges. A covalent bond forms when atoms share their electrons.

These chemical bonds form because electrons are gained or lost by atoms or shared between atoms. The electrons that make these chemical bonds are the valence electrons. For the elements we are considering, valence electrons are the electrons in an unfilled outermost energy level, or shell, of an atom.

(3) valence electrons, rule of 8

For a basic understanding of chemical reactions, we will look at valence electrons and how they affect whether two elements might chemically react to form a compound, whether by ionic or covalent bonds.

A general principle is that an atom outside the first period is most stable when the atom has eight electrons in its outermost shell; for first period elements, this stability occurs with two electrons. A neutral atom in group 18 by itself has eight electrons in its outermost shell (or two in the case of helium), so it tends not to react with another atom. For typical chemical reactions, atoms with less than eight outermost shell electrons are needed.

Chemical reactions tend to satisfy what is called the rule of 8, or octet rule, by emptying out, adding to, or sharing electrons in an atom's normally outermost energy level. In some cases, an atom empties its original outermost energy level, leaving this donor atom with its next lower energy level as its new, full outermost energy level. In other cases, an atom tends to take or share enough electrons from another atom (or atoms) to get to eight electrons in the taker's/sharer's outermost shell (or two in the case of hydrogen, as illustrated in the next paragraph) – thus the name "rule of 8," with hydrogen and helium being the exceptions because their single shells are full with two electrons.

In covalent bonding, the bonded atoms share valence electrons so that each outermost energy level is full when the atom's own electrons and the shared electrons are added together. For example, in H_2O each hydrogen atom has one valence electron that it shares with the oxygen atom, which has six valence electrons of its own. These two shared electrons plus oxygen's own six electrons give oxygen eight in its outermost energy level. At the same time, each hydrogen atom shares a respective one of the oxygen atom's valence electrons, thereby giving each hydrogen atom two electrons in its one principal energy level.

(4) chemical reaction procedure

We will use the rule of 8 to work some chemical reaction equations. Because this rule applies regardless of which type of bond is formed, we will not concern ourselves with whether a particular reaction forms ionic or covalent bonds; however, we will use the ionic terms "givers" and "takers" rather than the covalent term "sharers."

To work the chemical reaction equations we will be doing, follow this procedure:

1.   Determine whether there are both a giver element and a taker element. For our purposes we will assume a giver comes only from group 1, 2, or 13 (thus having one, two, or three valence electrons to give away) and a taker comes only from group 15, 16, or 17 (having five, six, or seven valence electrons, and thus wanting to take three, two, or one more electrons, respectively). If we don't have both a giver and a taker element in a reaction equation that we use, write "no reaction."

2.   If there are a giver and a taker in the chemical reaction equation, assume that each giver atom gives all its valence electrons and that each taker atom takes only enough electrons to

have a total of eight electrons in its outermost shell (or two in the case of hydrogen). Determine how many atoms of each element are needed so that the number of given valence electrons and the number of taken valence electrons are equal.

    3.   Use the proper chemical symbols to write the compound. Put the giver element first, the taker element second, and use subscript numbers to designate how many atoms of each element are needed (do not put any subscript if only one atom is needed).

    4.   Once you have figured out the compound that is made if there is a reaction, you then need to balance both sides of the equation. That is, the number of atoms for a given element in the reaction has to be the same on both sides of the equation. For example, to form Na_2O, you need to start with two sodium atoms and one oxygen atom: $2Na + O \rightarrow Na_2O$. This balancing satisfies the law of conservation of mass. This law says that matter cannot be either created or destroyed in ordinary chemical or physical changes (remember, we are dealing with chemical reactions, not nuclear ones like when we were talking about nucleosynthesis).

(5) three types of chemical reactions

From our definitions, we now know that a compound is a substance made of two or more different types of atoms, or elements, combined in definite proportions (for example, H_2O). These atoms are held together by a chemical bond to form the compound. We have briefly considered ionic bonds and covalent bonds.

In the reactions we are considering, a chemical reaction causes a change such that one or more elements or compounds (the reactants) form one or more new compounds* (the products). When written in a chemical reaction equation, the reactants are to the left of the arrow and the products are to the right of the arrow. We will look at examples below as we consider three different types of chemical reactions. [*Sometimes a chemical reaction occurs between atoms of a single element to form a molecule of the same element, such as two atoms of hydrogen bonding to form a hydrogen molecule, H_2. We are only looking at reactions that make compounds having different elements.]

(a) synthesis reactions

In synthesis reactions, two or more reactants combine to make one more-complex product. Examples of this type of reaction include:

    $2H_2 + O_2 \rightarrow 2H_2O$

    $2Na + Cl_2 \rightarrow 2NaCl$

$$CO_2 + H_2O \rightarrow H_2CO_3$$

(b) single replacement reactions

In a single replacement reaction (also called a single displacement reaction), one element replaces, or displaces, another element that was part of a compound, thereby forming a new compound. Specific examples include:

$$Cl_2 + 2NaBr \rightarrow 2NaCl + Br_2$$

$$Zn + CuCl_2 \rightarrow ZnCl_2 + Cu$$

$$Mg + 2HCl \rightarrow MgCl_2 + H_2$$

(c) double replacement reactions

In a double replacement reaction, two compounds switch partners to form two new compounds. For example –

$$2KOH + H_2SO_4 \rightarrow K_2SO_4 + 2H_2O$$

$$FeS + 2HCl \rightarrow FeCl_2 + H_2S$$

$$AgNO_3 + NaCl \rightarrow AgCl + NaNO_3$$

(6) acids, bases, and salts

Acids, bases and salts are three important types of compounds.

(a) one definition of acids and bases

The terms "acid" and "base" can each be defined in different ways. According to one definition: a compound is an acid if it produces H^+ ions when the compound is put in water, and a compound is a base if it produces OH^- ions when the compound is put in water.

(b) detecting acids and bases

Simple tests for detecting acids and bases include using chemically coated indicator paper.

(1') litmus test

One way to test for acids and bases is to run a litmus test. A type of chemically coated indicator paper called litmus paper reacts to acids and bases in different ways.

One type of litmus paper, referred to as red litmus paper, stays its original color when moistened with an acid. If red litmus paper contacts a base, however, the paper turns a bluish color.

On the other hand, blue litmus paper turns a reddish color in the presence of an acid. It does not change color when testing a base.

(2') pH

Another type of indicator paper can change color to indicate the pH level of a substance. pH is a numerical measurement of acidity or alkalinity (base-ness) of an acid or base. The term "pH" stands for "potential of hydrogen" (see the definition above referring to H^+ ions, for example).

A substance with a pH of 7 is neither an acid nor a base, it's neutral. A substance with a pH less than 7 is an acid. A substance with a pH greater than 7 is a base.

The pH scale is a logarithmic scale, which means it goes up or down in increments of powers of 10. For example, an acid with pH of 4 is 10 times more acidic than an acid of pH 5; pH of 3 is 100 times more acidic than pH 5; and pH of 2 is 1,000 times more acidic than pH 5. The lower the pH number for an acid, the stronger the acid; for bases, the larger the pH number, the stronger the base.

(c) salts

When an H^+ acid and an OH^- base are mixed in proper proportions, a neutralization reaction occurs. The result is that neither an acid nor a base remains; instead, water and a compound called a salt are formed. For example, in the following we have an acid + a base producing water + a salt: HCl (aq) + NaOH (aq) → HOH (l) + NaCl (aq)

2. mixtures

We have just finished learning about pure substances, which include elements and compounds. We will now look at the other major class of matter, mixtures.

A mixture is a combination of two or more distinct chemical substances that keep their own characteristics in the mixture. The different substances that make up the mixture can be separated by physical techniques; they are not chemically bonded with each other. For example, if you mix some salt in a glass of water so all the salt dissolves in the water, you have a mixture in which the water still has its properties of water and the salt still has its properties of salt, and you can separate the two by heating the mixture to evaporate the water and leave the salt as a solid residue.

There are two groups of mixtures: homogeneous mixtures and heterogeneous mixtures.

a. homogeneous mixtures: solutions

A homogeneous mixture, or solution, is uniform throughout a given sample of the mixture, but it is not necessarily uniform with other samples. For example, if you make some tea

25

in a glass, the solution of tea and water in that glass will be the same throughout that volume. Any spoonful of tea from that glass should be the same as any other spoonful from that same glass. But if you make another glass of tea, the ratio of tea to water will likely not be exactly the same as the ratio of tea to water in the first glass of tea that you made.

A solution includes a solvent, which is the medium in which another substance dissolves. Water is sometimes called the universal solvent because water dissolves many substances. The other substance, the one that dissolves in a solvent to form a solution, is called the solute. For example, fruit drink powder is a solute that is to be mixed with water (the solvent) to make the liquid fruit drink, which is a solution. Food coloring (the solute) mixed with water (the solvent) is another example of a solution. The amounts of solute and solvent in a solution relate to the concepts of concentration and dilution.

Concentration refers to the strength of a solution. One way to express concentration is to divide the quantity of one substance (for example, a solute) by the quantity of another substance that the first substance is mixed with (for example, a solvent). To illustrate: the concentration of 20 grams of fruit drink powder dissolved in 100 milliliters of water is 20 g/100 mL, or .2 g/mL.

Dilution refers to the process of reducing the concentration of one substance relative to another substance in a solution. For example, the concentration of a solute dissolved in water is diluted by adding more water. Referring to the example in the previous paragraph, the concentration is diluted to .1 g/mL if another 100 mL of water is added to the mixture. This is because the solution would then have the original 20 grams of fruit drink powder mixed in 200 milliliters of water (the original 100 milliliters plus the additional 100 milliliters).

b. heterogeneous mixtures: colloids and suspensions

A heterogeneous mixture is not uniform even within a given sample of the mixture (and clearly not from one sample to another sample). We will consider two types of heterogeneous mixtures: colloids and suspensions.

A colloid has a substance, called the dispersed substance, distributed in another substance, called the continuous or dispersing substance. For example, milk is a colloid made of fat deposits (the dispersed substance) non-uniformly distributed in the rest of the milk (the dispersing substance) – these separate into curds and whey when milk is mixed with vinegar, for example. Other examples of colloids include: fog, smoke, shaving cream, mayonnaise, and marshmallows.

A suspension is a heterogeneous mixture having particles that settle out of the mixture within a relatively short period of time. For example, beans stay suspended in water as long as mixing energy is applied; but when the mixing is stopped, the beans fall out of suspension. Similarly, orange pulp falls out of suspension in orange juice shortly after you stop shaking the juice container. Another example is river water with dirt, sticks, and other particles that quickly settle out when the water no longer flows (as in a sample collected in a bucket).

IV. Energy in Our Everyday World – Potential, Kinetic, Wave (a little physics)

A. introduction

Now that we have learned some things about the ordinary kinds of matter we encounter every day, we will study some of the types of energy in our everyday world.

We will consider the potential energy and the kinetic energy of objects that have mass, such as books and roller coasters. These objects exist in three dimensions and are made of ordinary matter like we have just studied; however, we will model them as solid objects and draw them as dots. The dot represents all the mass of the object concentrated at a single location in space.

After considering potential energy and kinetic energy of objects made of ordinary matter, we will learn about waves and the energy they transfer.

B. potential energy and kinetic energy

1. background

Before we get to potential energy and kinetic energy themselves, there's some background we need.

We will begin with motion of an object through space. When an object moves through space, the object changes its position in space. This change of position is measured as distance. Moving over this distance occurs during some amount of time. Dividing the distance traveled by the total time it took the object to move that distance gives us speed.

When speed changes, we have acceleration. If an object that has mass accelerates, a net non-zero total force must be acting on the object (the term "net non-zero total force" means that there could be multiple forces acting on the object and when the forces are added up, their sum is not zero; that is, there is some resulting push or pull on the object in some direction).

Force acting over a distance can produce work.

As we study this background information, we will also come across Newton's first and second laws of motion.

With this background, we will then consider potential energy and kinetic energy as well as the law of conservation of energy.

2. motion

a. space and time

In Newtonian physics (also called classical physics), space and time are separate properties of our universe. In relativistic physics (based on Einstein's theories of special and general relativity), space and time are combined as space-time (or spacetime). We will consider space and time in the Newtonian or classical physics sense. As such –

Space is a property of our universe in which something exists. It is the where or how far of existence. "Where" refers to a position or location in space. "How far" implies two or more locations in space and some measurement of the separation between them.

Time is a property of our universe during which something exists. It is the when or how long of existence. "When" refers to a particular point in time. "How long" implies a change between two or more instants in time, such as the time that elapses between two specific but different "whens."

b. speed and direction, velocity

(1) speed

When an object moves through space, it changes its position. If the change in position begins at point A and ends at point B, we call a measurement of the space between A and B distance. If point A corresponds to a point $x_{initial}$ relative to the origin of an x-axis that we define through both A and B, and if point B corresponds to a point x_{final} from the origin of the x-axis, the distance between A and B is $x_{final} - x_{initial}$. A shorthand symbol for $x_{final} - x_{initial}$ is Δx. The Δ symbol is the Greek letter delta, and we use it to mean change or difference (i.e., a subtraction relationship).

The change in position referred to in the previous paragraph occurs during some amount of time. As the object departs from point A, someone marks the time; we can call this $t_{initial}$. When the object arrives at point B, the arrival time is noted; we can call this t_{final}. The time it takes to go from point A to point B is then $t_{final} - t_{initial}$, or Δt.

If we measure change in position over a very short period of time, an instant in time, we have the instantaneous speed. You can think of the speedometer on a car as showing the instantaneous speed, the "how fast are we going" at the instant we look at the speedometer.

If we measure the change in position that occurs over a longer period of time, we call the speed the average speed. To find average speed, we determine the total distance traveled (Δx) and the total time it took us to travel that distance (Δt) (this includes any stops, when we had an instantaneous speed of 0!). It is this average speed that we will work with, which we write as average speed $= \Delta x / \Delta t$. The general units of speed are distance/time.

(2) direction

We can use average speed to determine how far an object has moved in a given period of time (rearranging the above equation: average speed · Δt = Δx). So, speed is an important characteristic of the motion of an object. But is that enough for us to know where that object is in space as a result of its speed? No, we also need to know the direction(s) of the object's motion.

(3) velocity

When we know both speed and direction of a moving object, we know the velocity of the object. That is, velocity includes both speed and direction. Speed tells us the rate of change in position of the object, and direction tells us which way in space that change is occurring; both of these together gives us velocity.

(4) Newton's first law of motion

Sir Isaac Newton's first law of motion, which is also known as the law of inertia, says:

A body in constant motion (or at rest) stays in the same constant motion (or at rest) unless a net external force acts on the object.

Newton's first law refers to constant motion. For our purposes in thinking about this law, let's substitute the word "velocity" for the word "motion." Since velocity includes both speed and direction, this law tells us that an object in constant motion moves at an unchanging speed in an unchanging direction. This motion at a constant speed in a straight line will continue forever unless some net force causes a change in either speed or direction (or both).

c. acceleration and force

(1) acceleration

Acceleration is the rate of change of velocity. If there is a change in speed or a change in direction or changes in both, we have acceleration. We will consider only changes in speed.

Looking just at speed, average acceleration is the change in speed divided by the time over which the speed change occurs: average acceleration = (change in speed)/(total time over which the speed changes) = Δ speed / Δ time = $(speed_{final} - speed_{initial}) / (time_{final} - time_{initial})$.

The general units of acceleration are distance/time^2 .

(2) Newton's second law of motion

Newton's second law of motion relates acceleration to force and mass. One statement of this law says:

The acceleration of an object of constant mass is directly proportional to the net force acting on it and inversely proportional to the mass of the object: acceleration = net force / mass

A more common form of this law is net force = mass · acceleration, or $F_{net} = ma$.

A common unit of force is (kilogram · meters)/seconds^2 or (kg · m) / s^2 . This combination of units is called a Newton, abbreviated N.

So, force acting on an object that has mass is related to acceleration of the object, which acceleration is related to change in the object's speed, which speed is related to change in position of the object. There is no acceleration and no net force acting on the object if there is no change in direction and no change in speed as the object changes its position. If the speed or direction does change, there must be acceleration and a net force on the object.

3. work and energy

Let's define a system consisting of Earth, a book lying on the ground at the surface of Earth, and a table on the ground near the book. The table top, supported by the table's legs, is some vertical distance above the ground. Earth provides a source of gravity. If I (who am not part of the defined system) lift the book vertically from the ground to the table top, I have done work on the system. For our purposes, we will consider only a force that moves an object in the same direction of the force and in a straight line. In our example, I exert an upward force on the book to move it vertically up from the ground to the top of the table. The work I have done in lifting the book is calculated by multiplying (A) the gravitational force acting on the book along the direction the book is moved by (B) the change in position of the book (i.e., the distance, d, that I lift it). From Newton's second law, the gravitational force, $F_{gravity}$, acting on the book equals the mass of the book (m_{book}) multiplied by gravitational acceleration (g). So, the work, W, as defined above is W = gravitational force on book x change in vertical position of book = $F_{gravity}$ x distance = $[(m_{book})g]$ x d = $(m_{book})gd$. This amount of work that I have done lifting the book has changed the energy of the system. The change in energy equals the work done on the system. If we determine work, we determine a change in energy.

There are many forms of energy: thermal energy, chemical energy, potential energy, kinetic energy, to name a few. Energy can be added to a system, removed from the system, and changed from one form to another (e.g., from potential energy to kinetic energy); however, all of the energy has to be accounted for. This relates to what is called the law of conservation of energy. Keeping track of all the forms of energy can be difficult, so sometimes we simplify or idealize the system so it is easier to work with the main energy components of interest; however, we still need to account for what we are considering in the simplified system.

In class we will work with a system that includes Earth, a roller coaster car, and the roller coaster track. We will simplify the system by disregarding friction between the car and track, wind resistance, and any other forms of energy besides potential energy and kinetic energy. Work will be done on the system whereby there is a change in potential energy (the roller coaster car is lifted from a reference height to the top of the first hill of the roller coaster track), and then an event will occur in which we account for changes in potential energy (PE) and kinetic energy (KE) within the system (the car races down the track from the hilltop to the bottom, which is at the reference height). To comply with the conservation of energy, the mathematical equation representing this is $\Delta KE + \Delta PE = 0$. Remembering that the Δ symbol means change or difference, we will analyze this equation by expanding it to the following equivalent equation:

$$(KE_{final} - KE_{initial}) + (PE_{final} - PE_{initial}) = 0$$

We will consider the initial potential energy in this system in the context of the concept of work and gravitational force described above. The kinetic energy we will consider relates to translational motion, which is movement of an object from one location in space to another location (as opposed to rotational movement, such as a top spinning on its axis at a single place in space). For this translational motion, kinetic energy = $\frac{1}{2} (mv^2)$, where m is the mass of the object moving through space (the roller coaster car in our system) and v is that object's velocity. Although we will use the letter v for velocity, kinetic energy just depends on the speed part of that term.

    C. wave energy

        1. definitions: waves, mechanical and electromagnetic

Let's say that you want to say something to your friend as both of you walk down the hall between classes. Once you've thought what to say, your body's energy forces air from your lungs past your vocal cords to make them vibrate. These vibrations pass from particle to particle in your mouth and then outside of your mouth to produce a sound wave in the air. If the sound wave reaches your friend's ear, the energy transferred by the sound wave vibrates the parts inside your friend's ear so that she hears the sound (if the vibrations get converted to nerve impulses).

The vibrations of the sound wave momentarily displace something as the wave travels, but the vibrations do not move that something from where the wave began to where the wave ends up. In talking with your friend, when the sound wave travels from your mouth to your friend's ears, the energy of that sound wave vibrates air molecules between your mouth and your friend's ears; however, the energy of the wave does not move air molecules from your mouth into your friend's ears. So, a wave is a disturbance that transfers energy from place to place. Some source of energy provides the disturbance, a vibration; in the case of you talking to your friend, the energy comes from your body and vibrates your vocal cords, which then vibrate air molecules.

A mechanical wave is one kind of wave, and a sound wave is one type of mechanical wave. A mechanical wave vibrates particles of matter, which we call a medium. To exist, a mechanical wave requires a medium made of matter. In our example about talking to your friend, the medium between you two includes the air molecules in the space around you. The matter passes vibrations from particle to particle, thereby transferring the wave energy; but the particles are not transported from one location to another, such as from where the wave begins to where it ends. Because a mechanical wave cannot exist without a medium, a mechanical wave cannot exist in outer space where there is not enough matter for vibrations to pass from one particle to another.

An electromagnetic wave, on the other hand, does not need a material medium to exist. Instead, an electromagnetic wave is made of vibrating electric and magnetic fields, so an electromagnetic wave can exist in outer space. An electromagnetic wave can also travel through some mediums, but the electromagnetic wave does not use the medium to exist; however, the speed and direction of the electromagnetic wave can be affected by the medium. Visible light is one example of an electromagnetic wave.

2. form of a wave

The direction in which the particles of the medium for a mechanical wave or the fields of an electromagnetic wave vibrate defines the form of the wave.

Vibration perpendicular to the direction that the wave travels (i.e., the direction in which energy is transferred) characterizes a transverse wave. Think of a rope lying on a table and think of making a wave on that rope by wiggling one end of the rope back and forth along the table top. Energy moves from the wiggled end to the other end along the length of the rope; however, the rope (each small segment of it) moves perpendicularly to this direction of energy transfer.

Vibration parallel to the direction that the wave travels characterizes a longitudinal wave. Think of a coiled spring toy stretched out along the length of a table. What will happen if you push and pull one end of the spring? Here, energy will move from the pushed/pulled end of the spring to the other end. The coils of the spring move parallel to this energy transfer (or wave) direction.

Some mechanical waves are transverse waves and some are longitudinal waves. Electromagnetic waves in free space have a transverse form.

3. properties of waves

Whether a wave is a mechanical wave or an electromagnetic wave, or whether it is longitudinal or transverse, waves have the following properties: amplitude, wavelength, frequency, and speed.

a. amplitude

The amplitude of a mechanical wave is the maximum distance the particles of the medium move away from their rest positions as the energy passes through the medium. For an electromagnetic wave, amplitude relates to the magnitude of the electric and magnetic fields that make up the wave.

b. wavelength

The wavelength of a wave is the distance between two corresponding points of the wave, which is the distance across one complete oscillation (cycle of vibration) of the wave. So, the units of wavelength are distance/oscillation, but we typically just state the distance (e.g., a wavelength of 3 cm, without reference to the "per oscillation").

c. frequency

The frequency of a wave is the number of oscillations of the wave that pass a given point in a certain amount of time.

To calculate frequency, count the number of oscillations over some period of time and then divide that number of oscillations by the time. For example, if you count 10 oscillations of a wave passing some reference point in 5 seconds, the frequency is 10 oscillations/5 seconds = 2 oscillations/second = 2 hertz. The unit "hertz" (abbreviated Hz) means oscillations/second, or more commonly, cycles/second; however, in units, hertz simply means 1/second.

d. speed

Here is the equation for the speed of a wave: speed = wavelength x frequency

From our previous study of speed, we know that the units of speed should be distance/time. If we have a wave with a wavelength of 3 cm and a frequency of 250 hertz, the wave has a speed of 3 cm x 250 hertz = 3 cm x 250/s = 750 cm/s , which are speed units of distance/time.

e. effect of motion between source and receiver of a wave

Assume that you are standing along a street and that a siren on a fire truck produces sound at a single frequency. The fire truck with its siren on speeds by you. The frequencies of the siren sounds you hear are different from the single frequency that the siren emits as it approaches you and then moves away from you.

(1) Doppler effect

If the source of a wave and a receiver of the wave move relative to each other through space, the receiver detects a different frequency than was generated by the source. This is called the Doppler effect. The example about the fire truck siren and you illustrates this effect. This effect applies to mechanical waves (e.g., sound) and electromagnetic waves (e.g., visible light). As we will see later, the Doppler effect is an important tool that astronomers use to learn about objects in outer space.

(2) expansion of universe

A similar result occurs for an electromagnetic wave from far away in outer space because of the expansion of the universe through which the electromagnetic wave travels. As a result of this expansion, the light wave we receive from the distant source has a different wavelength (and frequency) than that light wave had at its source. This is another important tool that astronomers use to learn about objects in outer space.

4. wave interactions

Waves, whether mechanical or electromagnetic and regardless of their form, interact with the mediums they travel through, with objects in the medium, and with other waves. These interactions include reflection, refraction, diffraction, and interference. But first, how do we represent waves in drawings?

a. wave diagrams

We can draw the waveform of a wave. For example, we can draw the shape of the transverse wave on a wiggled rope lying on a table at an instant in time.

We can also represent the wavefront of a wave, either as a single straight or curved line or a series of straight or curved lines. For example, think of these as parts of ripples formed when you drop a pebble on a pond.

We can also draw a ray diagram (an arrow) to represent the direction of a wave.

Now, let's move on to the wave interactions of reflection, refraction, diffraction, and interference.

b. reflection

Reflection occurs when a wave encounters a boundary and "bounces off" of it; that is, the flow of energy is redirected away from the boundary. When this happens, we say the wave has reflected off the boundary. When the wave strikes the boundary at an angle, it is reflected at the same angle. This is called the law of reflection and is stated: angle of reflection = angle of incidence. These angles are measured from a reference line that is perpendicular to the reflection surface at the point of reflection. We will use ray diagrams to represent reflection.

c. refraction

Refraction of a wave occurs when the wave changes direction as it passes from one medium into another medium. For this to happen, the wave must enter the second medium at an oblique angle and the speed of the wave must change. We will also use ray diagrams to represent refraction.

d. diffraction

As a wave moves in a medium, it may spread out or bend within the medium as it passes through an opening or around the edge of a barrier in the medium. This spreading or bending is called diffraction. This noticeably occurs if the width of the opening is comparable to the wavelength of the wave. We will use wavefront diagrams to illustrate diffraction.

e. interference

When two or more waves are in the same space at the same time, they interfere. When interference occurs, the amplitudes of the individual waves add or subtract depending on the alignment of the waves. We will use waveform diagrams to illustrate interference.

5. electromagnetic waves

We live in a universe filled with electromagnetic waves. Our sun and other stars constantly generate electromagnetic waves. Electric lights in our homes, microwave ovens, cancer radiation treatments, satellite communications such as GPS and satellite radio and television signals, cell phones, WiFi, and television and radio signals broadcast from tall antennas all generate or use electromagnetic waves. Another example is the cosmic microwave background radiation generated from the Big Bang. Some of these waves are passing around and through you as you read this (remember, electromagnetic waves do not need a medium to exist but they can pass through and interact with a medium).

We have been using visible light, which consists of electromagnetic waves, to study properties and interactions of waves in general and so electromagnetic waves have the properties we have already studied (amplitude, wavelength, frequency, and speed). Electromagnetic waves also interact in ways we have studied (reflect, refract, diffract and interfere). However, the nature and study of electromagnetic waves have led to deeper understandings of our universe and the nature of both waves and matter.

a. nature of electromagnetic waves

As with all waves, electromagnetic waves transfer energy. The disturbances that transfer the energy are vibrations or oscillations in an electric field and a magnetic field. The changing electric field and the changing magnetic field are perpendicular to each other in free space. These fields vibrate transversely to the direction that the wave transfers the energy in free space, so these electromagnetic waves are transverse waves.

Because these fields cyclically vibrate or oscillate, these changing electric and magnetic fields keep generating each other as described by Maxwell's equations, which we will consider next.

(1) Maxwell's equations

In the mid-1860s James Clerk Maxwell unified our understanding of electricity and magnetism. He brought together discoveries by others, added something himself, and the result is now set forth in four equations relating to electric charges, magnetic charges, electric fields and magnetic fields.

(a) electric charges, electric fields

The first equation, known as Gauss's law of electricity, tells us that electric charges produce electric fields. Have you ever demonstrated this by loading a balloon with electric charges and using that charged balloon to move pieces of paper (or your hair!) without touching them? The electric charges on the balloon create an electric field extending beyond the balloon.

(b) magnetic charges, magnetic fields

Using a magnet you can attract or repel magnetic objects from a distance because of the magnetic field that extends from the magnet. You may have done this using a bar magnet from which a magnetic field is produced. Such a bar magnet has both a north pole and a south pole. A magnet with only a north pole or only a south pole has never been found. That is, north and south poles always exist together; there is no separate north-pole-only magnet or south-pole-only magnet. Stated another way, no magnetic monopoles have been found. The second equation of Maxwell's four, known as Gauss's law of magnetism, relates to this observation.

(c) changing magnetic fields produce electric fields

A changing magnetic field produces an induced electric field, which can be used to move electrons in a wire to generate electricity (the field can change, or the wire can be moved relative to a constant field). This is the third equation, known as Faraday's law of induction. This is the basis of electric generators.

(d) changing electric fields produce magnetic fields

The fourth equation, known as Ampere's law, says that a moving electric charge (e.g., an electric current in a wire) produces a magnetic field. The fourth equation also says that a changing electric field generates a magnetic field (this is Maxwell's addition). It is this latter feature of a changing electric field producing a magnetic field which is critical to Maxwell's equations defining electromagnetic waves.

(e) so what?

So what do we learn from Maxwell's equations that is relevant to electromagnetic waves? Well, look back at the titles of sections (c) and (d) above and think about what happens if the produced electric fields and magnetic fields are also changing: each keeps the other going!

(2) waves or particles?

So Maxwell showed us that light and other electromagnetic energy (any of which we call electromagnetic radiation) travel as electromagnetic waves. But this is only half of the story (now it's time for some weird quantum physics stuff!).

Since at least the time of Newton, scientists argued about whether visible light (and after Maxwell's time, any electromagnetic radiation) was a wave or a stream of particles. People arguing for waves pointed to observations about diffraction and interference (which we have already studied) and polarization to support their position. Also, Maxwell's work seemed to establish once and for all that electromagnetic radiation is transferred as a wave (of course, nothing in science is established once and for all; some later discovery might come along and show something we thought was correct is not – as has happened throughout history).

However, even after Maxwell's work, scientists could not explain something called the photoelectric effect if light travels as a wave. Einstein finally explained this effect by characterizing light as a stream of discrete units of energy. However, note that these discrete units, referred to as particles, are not particles of matter, they are massless particles called photons. Although Einstein's explanation solved the photoelectric effect issue, his stream of particles explanation did not explain the experiments where electromagnetic radiation acted as a wave.

So, sometimes an electromagnetic wave acts as a wave, and sometimes it acts as a stream of particles. Because of this duality (sometimes acting as a wave, other times acting as a stream of particles), and because we have no better mental imagery or language, the term "wave-particle duality" is used to describe the nature of electromagnetic radiation.

(a) polarization

According to Maxwell's equations, an electromagnetic wave consists of a vibrating electric field and a vibrating magnetic field which are intertwined and perpendicular to each other. But what about the relationship between these fields and a narrow slit in some material that the electromagnetic wave is focused on?

Think of a beam of white light shining on a thin piece of plastic that has a narrow slit in it. Only light waves having their electric fields aligned with the slit will pass through – the light has been polarized. If we take a second piece of plastic with a narrow slit and put it at 45° to the first one and then take a third piece of plastic with a similar slit in it and place it over the first two pieces so that the third slit is perpendicular to the slit in the first sheet, no light passes through because no matter the polarization of the electric field of the light, it can't pass through all three of the different orientations of slits. According to classical physics, this can be most easily explained if the light (electromagnetic radiation) is a wave rather than a stream of particles.

(b) photoelectric effect

In the late 1800s Max Planck worked on the problem of explaining energy that is emitted by objects in the form of electromagnetic radiation related to the temperature of the object. It was experimentally known that at a given temperature, an object emitted electromagnetic radiation at certain frequencies and intensities. To mathematically describe these experimental results, Planck discovered that he had to assume there is some smallest amount ("quantum") of energy and that the detected energies were multiples of this smallest amount. In 1900 Planck published a paper explaining his discovery that energy cannot be continuously changed in any amount, but only in very small increments.

Prior to Planck's work, scientists knew that sometimes when light shines on particular types of matter (metals in particular), electrons are ejected. This is the photoelectric effect. However, some light would produce this effect, while other light, no matter how bright or how long it was shined on the photoelectric material, did not produce this effect. These observations were contrary to what should happen if light is a wave.

In 1905 Einstein explained the photoelectric effect by using Planck's idea of the quantum. Einstein proposed that light must be a stream of discrete units of energy (now called photons) and if one of these photons had the right amount of energy, it would knock one electron free from the photoelectric material. Each photon would knock one electron free, thereby producing electricity in a circuit connected to the photoelectric material (i.e., a flow of electrons, or electric current outside the photoelectric material).

(3) energy of a photon vs. energy of a wave

Because any flow of electromagnetic radiation has both wave and particle characteristics, how do we measure the energy? We look at both the energy of a photon (i.e., a massless particle of electromagnetic radiation) and the energy of the wave.

The energy of a photon is proportional to its frequency. Specifically, the energy of a photon (E_{photon}) equals a number known as Planck's constant (h) times frequency (f): $E_{photon} = hf$. Planck's constant, which relates to the smallest quantum of energy, has a particular value, namely, $h \approx 6.626 \times 10^{-34}$ Joule-seconds.

As for the energy of the overall wave, any wave (mechanical or electromagnetic) has energy, or intensity, which is proportional to the amplitude squared: $E_{wave} \propto A^2$.

For electromagnetic radiation, the total energy relates to the amplitude of the wave as just mentioned ($E_{wave} \alpha A^2$). The amplitude of the electromagnetic wave is related to the number of photons in the wave – the more photons there are, the greater the amplitude. Thus, a brighter (higher amplitude) light is brighter because it has more photons than a dimmer (lower amplitude) light has. On the other hand, each of the individual photons has a respective energy based on wave frequency for that particular photon. So, there is the energy of individual photons ($E = hf$), and there is the total energy (or intensity) of the overall electromagnetic wave related to how many photons are in the wave.

b. the electromagnetic spectrum

Whether we characterize electromagnetic radiation as a wave or as a particle, the range of this radiation is called the electromagnetic spectrum. The electromagnetic spectrum is based on the wavelengths and frequencies of these waves/photons. At one end of the spectrum we have the longest wavelength, lowest frequency electromagnetic radiation, and at the other end we have the shortest wavelength, highest frequency electromagnetic radiation. Thus, in terms of photon energy, we have the lowest energy at one end (the longest wavelength, lowest frequency end) and the highest energy at the other (the shortest wavelength, highest frequency end).

Although all the electromagnetic radiation throughout the entire spectrum is transferred in the same way, humans have put the different photon energies to different uses. As a result we have given different names to different ranges of the electromagnetic spectrum.

At the longest wavelength, lowest frequency (and thus lowest photon energy) end of the spectrum are the radio waves. These include the waves/photons used for television, radio, cell phone, and other communications. Sometimes included in, but often listed separately from, the radio wave portion of the spectrum are microwaves, which include the electromagnetic energies used in microwave ovens, for example.

At shorter wavelengths/higher frequencies than microwaves is the infrared portion of the electromagnetic spectrum. Infrared radiation is associated with heat and thus was part of what Planck studied when he discovered the quantum. The digital thermometer we have used in class detects infrared radiation. Because infrared radiation is relatively low energy and invisible to humans, some remote controls, such as for televisions, use infrared radiation to send their control signals.

Visible light has shorter wavelengths and higher frequencies than infrared radiation. Visible light is a narrow region of the overall electromagnetic spectrum, but it is critical to humans since this is the electromagnetic radiation that our eyes respond to, enabling us to see. The low energy end of visible light is red light. Going up from red light, the main segments of this portion of the spectrum include orange, yellow, green, blue and violet light, from which we have the acronym ROYGBV.

Ultraviolet light is above ("ultra") the violet end of the visible light portion of the spectrum, and so it is invisible to us. The ultraviolet portion of the electromagnetic spectrum includes the uva and uvb radiation that we guard against with suitable sunscreen, sunblock, and sunglasses.

X-rays and gamma rays, which are the remaining two portions of the electromagnetic spectrum, have even shorter wavelengths and higher frequencies/photon energies. These can be highly damaging to humans; but when used in controlled amounts, they have beneficial medical uses. For example, x-rays can be used to make images of dense structures (such as bones) inside our bodies, and gamma rays can be used in surgical procedures and cancer treatments.

c. generating electromagnetic waves

(1) generally

According to Maxwell's equations, electromagnetic waves have an electric field and a magnetic field changing over time, with each changing field producing the other type of field. These two related electric and magnetic fields move out from the source together, continually regenerating each other. This is the classical explanation based on Maxwell's equations.

If we consider photon production at a quantum physics level, however, one way to produce a stream of photons, and thus an electromagnetic wave, is to move electrons between energy levels. When an electron in an atom moves from a higher energy level to a lower energy level, energy must be released and this is done by emitting a photon having the appropriate energy.

(2) incandescence

An incandescent light bulb is the type of bulb that has been used in homes for decades, but current laws in the U.S. are causing this type of bulb to be phased out because it is not energy efficient enough. A typical incandescent bulb has a filament through which electricity flows when the bulb is on.

Electric current flowing through an incandescent light bulb causes the filament to get hot. As a result, electrons in the filament move to higher energy states and then emit photons as they drop to lower energies. Part of this produces infrared radiation, which we feel as an increase in temperature (the bulb gets hot!). Higher energy photons at visible light frequencies are also generated, thereby producing visible light. However, most of the energy produced through an incandescent bulb is outside the visible range and so that energy is not useful (unless you are trying to warm something up with the light bulb!).

(3) fluorescence and phosphorescence

Fluorescence and phosphorescence occur when photons having sufficient energy are absorbed by electrons in atoms of the fluorescent or phosphorescent material, whereby these electrons move to higher energy levels; when these excited electrons return to lower energy levels, photons are released. This produces visible (and possibly other) electromagnetic radiation.

For a specific example, let's consider a fluorescent light tube such as in the overhead lights in our classroom. When these lights are on, electric current flows through a gas inside the fluorescent light tube. This electric current energizes electrons in atoms of the gas to higher energy levels, which produces photons of ultraviolet electromagnetic radiation when the electrons return to lower energy levels. The ultraviolet photons strike a coating on the inside of the fluorescent tube. Electrons of this coating material absorb the energy of these photons and move to higher energy levels. When these electrons of the coating return to lower energy levels, photons of visible light are produced.

The difference between what we call fluorescence and what we call phosphorescence has to do with the time it takes excited (i.e., higher energy level) electrons to move to lower energy levels and release photons. If all the electron transitions occur virtually instantaneously so that light emission stops immediately when the power to the light is turned off, we call it fluorescence. If the transitions continue for some time afterwards, whereby light continues to be emitted for a while, we call it phosphorescence.

d. "seeing" electromagnetic waves

(1) within the visible range

Our eyes respond to electromagnetic waves having frequencies within the visible portion of the electromagnetic spectrum; or we could say that our eyes respond to photons having energies within this range according to E = hf. When our eyes receive electromagnetic radiation throughout this range of frequencies or energies all at one time, we perceive this collectively as white light. We can, however, use a prism or spectroscope to see and study individual components of this overall combination of waves or photon streams. The prism uses refraction to separate the frequencies or energies, and the spectroscopes we will use in class use diffraction to separate them.

(2) outside the visible range

Our eyes can't see electromagnetic radiation outside the visible portion of the spectrum, but equipment we build can detect this radiation by using photon-electron interactions. Photons of this electromagnetic radiation strike a material that is sensitive to the energies of those photons; this is used to enable electricity to flow.* Electrical circuits (such as in the form of a computer) convert this electricity into visible images. Because these created images can be made in colors within the visible range of the electromagnetic spectrum so that we can see them, when in fact the actual detected electromagnetic radiation is not at these visible frequencies, we call these images false color images. As we will discover later, false color images from detected radiation outside the visible range can be useful when studying the universe. [*This process can use the photoelectric effect described previously; however, most materials used today are semiconductors – in the photoelectric effect described by Einstein, metals are used and the electrons actually leave the metal whereas in a semiconductor, absorbed photons move electrons from states which permit little current to flow in the connected circuit to states where current can flow in the connected circuit.]

V. Using Electromagnetic Waves – A Key to Our Universe (a little astronomy)

A. introduction

When you look up at the sky at night, what do you see? What is coming to you from outer space that enables you to see those objects out there?

Much of what we learn about our universe beyond Earth comes from electromagnetic radiation generated from or reflected by the objects that are out there. One way to use this electromagnetic radiation is to separate it into its different frequencies. This can be done using spectroscopes. We will use spectroscopes to look at visible light spectra from incandescent and fluorescent lights (spectra is the plural of spectrum, and a spectrum shows the different frequencies in the received electromagnetic radiation). Astronomers and other scientists use spectra throughout the entire range of electromagnetic radiation to learn about our universe (recall the prior reference to false color images).

B. determining composition, direction, and speed of cosmic objects

As we have learned, matter can absorb and emit electromagnetic radiation. Absorption occurs with electrons absorbing photons and moving to higher energy levels, and emission occurs with electrons moving to lower energy levels and emitting photons. Spectroscopy uses a collection, or spectrum, of the absorbed or emitted electromagnetic radiation to determine quantitative and qualitative characteristics of this matter. These characteristics include the composition of an object (i.e., what elements it is made of), which direction an object is moving relative to us, and how fast it is moving in that direction. Before we look at these characteristics, what does a spectrum look like?

1. classes of spectra

There are two classes of spectra of electromagnetic radiation: continuous and discrete. As we consider these in terms of energy, remember this is photon energy which is related to wave frequency ($E_{photon} = hf$).

a. continuous spectra

In a continuous spectrum, we see a continuous range of electromagnetic energies. If these energies are within the visible range, we see them as different colors with one blending into the next along the continuous spectrum from red to violet (ROYGBV). Each of these colors corresponds to electromagnetic radiation having a respective frequency.

The generation of electromagnetic radiation by heating typically produces continuous spectra, such as what you observe through a spectroscope picking up light from an incandescent light bulb. The spectrum should appear continuous from red through violet.

b. discrete spectra: emission vs. absorption

In a discrete spectrum, we observe color lines or dark lines at specific photon energies or wave frequencies of electromagnetic radiation. Color lines occur at frequencies at which electron transitions emit photons; dark lines occur at frequencies at which electrons absorb photon energy.

Within the class of discrete spectra, an emission spectrum contains discrete (i.e., individual) color lines against a black background (i.e., there are black gaps between the color lines). We observe discrete emission spectra by looking through spectroscopes and picking up light from the overhead fluorescent lights in our classroom.

An absorption spectrum is another type of discrete spectrum. This type of spectrum appears as discrete dark lines against a continuous color spectrum background.

Spectra can be displayed graphically as well.

2. using spectroscopy

As mentioned above, we can use spectra of the electromagnetic radiation from objects in our universe to determine the chemical composition of the object, the direction the object is moving relative to Earth, and the speed the object is moving.

a. chemical composition

Each element of the Periodic Table has its own unique emission or absorption spectrum. A spectrum from an object (e.g., a star) can be analyzed by comparing the emission or absorption lines in the spectrum for that object with each of the known unique spectra of the elements in the Periodic Table. Where matches occur, we know those elements exist at the object being analyzed.

b. direction of movement

We can also use spectra of electromagnetic radiation coming from objects in outer space to determine their direction of movement. We use changes in wavelength or frequency to do this.

Remember the Doppler effect – the apparent change in frequency between what a source emits and what a receiver receives when there is relative motion between the source and the receiver. When there is relative movement away from each other (i.e., the distance separating the two is increasing), the wave the receiver receives has a lower frequency than the frequency the wave has at the source. The receiver receives a higher frequency than the frequency of the wave at the source when there is relative movement toward each other (i.e., the distance separating the two is decreasing). In either of these cases, for sound we say there is a change in pitch. But now we are dealing with light from an object in outer space.

With regard to visible light (and electromagnetic waves in general), rather than pitch, we refer to redshift and blueshift. When an object emits visible light and moves away from Earth, the wavelengths of the electromagnetic waves appear to us as lengthened, and so the frequencies we receive are lower. If frequencies of visible light shift lower, they are shifted toward the red (lower frequency, longer wavelength) end of the visible light range of the electromagnetic spectrum; so we have redshift. When an object emits visible light and moves toward Earth, the wavelengths of the individual waves we receive are shorter than their source wavelengths, and so the frequencies are higher. If frequencies of visible light shift higher, they are shifted toward the blue (higher frequency, shorter wavelength) end of the visible light range of the electromagnetic spectrum; so we have blueshift. Redshifts and blueshifts can occur because of either the Doppler effect or, for very distant objects, the expansion of the universe.

While we have stated the Doppler effect in terms of frequency, for light we will refer to wavelength instead of frequency. For electromagnetic waves we can talk in terms of either wavelength or frequency interchangeably because they are related by speed = wavelength x frequency, and speed is always constant at 3×10^8 meters/second (or, equivalently, 3×10^5 kilometers/second) for electromagnetic waves in outer space. Also, remember that wavelength, or frequency, changes observed for very distant objects can occur due to the expansion of our universe.

To determine redshift or blueshift, we look at a discrete spectrum of the received light. We compare the wavelengths of the received emission or absorption lines shown in the recorded spectrum to the wavelengths that those spectral lines would have if there were no relative movement between the source and us. From this information we can calculate redshift or blueshift (for which we use the letter "z") with the following equation:

$$z = (\lambda_{\text{received}} - \lambda_{\text{unshifted}}) / \lambda_{\text{unshifted}}$$

In this equation, $\lambda_{\text{received}}$ is the wavelength of the wave received from the object that is moving relative to us, and $\lambda_{\text{unshifted}}$ is the wavelength of the wave we would observe if there were no relative motion between its source and us (which is a value that can be determined in a laboratory on Earth).

If z is positive, the received wavelength $\lambda_{\text{received}}$ is larger (longer) than the unshifted value $\lambda_{\text{unshifted}}$, so the shift is toward the red, longer wavelength end of the visible spectrum. This lengthening of wavelength, which corresponds to lower frequency, means the object is moving away from us. If z is negative, the received wavelength $\lambda_{\text{received}}$ is smaller (shorter) than the unshifted value $\lambda_{\text{unshifted}}$, so the shift is toward the blue, shorter wavelength end of the visible spectrum. This shortening of wavelength, which corresponds to higher frequency, means the object is moving toward us.

c. speed of movement

To determine the speed that the object is moving toward or away from Earth, multiply z by the speed of light; therefore, speed $= z \cdot c$ (c stands for the speed light travels in free space). The positive or negative value of z tells the direction as explained above. For the speed of light, we will use the conventional but approximate value of 3×10^8 meters/second or 3×10^5 kilometers/second.

C. measuring distances in the universe

Astronomers have several methods for measuring distances from Earth to other objects in the universe. We will consider three of them: radar ranging, parallax, and Hubble's Law. But first, what units of measurement do we use?

1. units of distance

a. for distances within our solar system

Our basic unit of distance is the meter. For measurements beyond Earth, however, we tend to use at least kilometers (1 km = 1,000 meters = 10^3 meters). This is still a short unit of measurement for cosmic distances. So, within our solar system another unit for measuring distances is the astronomical unit, or AU, which is the average distance between Earth and the sun. An approximate value we will use for 1 AU is 150,000,000 kilometers = 1.5×10^8 km.

b. for distances beyond our solar system

Objects beyond our solar system are significantly farther away so that even the AU is too short of a unit. Instead, we use the light-year, which is the distance light travels in a year. One light-year equals approximately 9,500,000,000,000 km, or 9.5×10^{12} km (which is sometimes rounded further to 10 trillion kilometers, or 10^{13} km).

Another unit that you might see is the parsec (its abbreviation is pc). One parsec equals about 3.3 light-years.

Now let's see how we can measure distances and use these units.

2. radar ranging

Radar ranging is one way to measure distances from Earth to objects within our solar system. With radar ranging a radio wave, which is an electromagnetic wave having a wavelength in the longer wavelength, lower frequency radio wave portion of the electromagnetic spectrum, is sent from Earth to the object and the time is measured from when the wave was sent to when its reflection is received back on Earth. The measured time and the speed of the radio wave are plugged into our basic equation for speed, namely, speed = distance/time, which is then solved for distance. The speed value used here is the speed of the radio wave. Because this radio wave is an electromagnetic wave traveling mostly through outer space, we use the speed of any electromagnetic wave in free space, namely c, which is approximately 3×10^8 meters/second or 3×10^5 kilometers/second.

The distance calculated this way is the round-trip distance to the object and back. So, divide that distance by two to find the one-way distance between Earth and the object.

3. parallax

We can use parallax to measure distances from Earth to celestial objects out to a few thousand light-years from Earth. Parallax is the apparent change in position of an object when you look at it from different places. To demonstrate parallax, hold one of your arms out in front of you with your thumb pointing up, close one eye, and with your other (open) eye sight along your thumb to a vertical edge or line of a wall; then close the open eye and open the other eye. What do you notice? What if you move your thumb different distances from your eyes? Does your thumb appear to move? Does your thumb appear to move more the closer it is to your eyes?

We can use this apparent change in the position of a distant object to determine how far away it is. The object (e.g., a star) is observed from two different locations; this defines a baseline between the two observation points and it also defines two lines of sight from the observation points to the object. The baseline and the two lines of sight form a triangle. We can measure the distance between the two observation locations and thus determine the length of the baseline, so we know the length of one side of the triangle. For objects out in space, astronomers use the baseline defined between the positions of Earth six months apart in its orbit around the sun. The angle of the apparent change in position of the object at the intersection of the two sight lines can also be measured. From all this we have an isosceles triangle with a 2-AU long baseline and a measured apex angle determined from the observed parallax.

The distance to the object can be calculated with the following equation: distance = baseline x 57.3°/angle. The units for the angle should be degrees. For an astronomical object with which this distance-measuring technique is used, the angle is quite small (a fraction of a degree) because the object is so far away. These angles are typically given in arc minutes or arc seconds. One arc minute is 1/60 of 1°. One arc second is 1/3600 of 1°. For example, an angle measured as 7 arc minutes would be (7/60)° and an angle of 7 arc seconds would be (7/3600)°.

The distance calculated by the foregoing is the object-sun distance, but this is a close approximation of the object-Earth distance due to the relatively short baseline (in AUs) and the relatively long distance to the object (in light-years or parsecs – for further study look up where the term parsec comes from).

4. Hubble's Law

A third measurement technique returns us to Hubble's law: $v = H_0 d$, or speed $= H_0 \cdot$ distance.

We now know that we can determine speed of a celestial object by first using spectra of electromagnetic radiation to determine the redshift value z and by then multiplying that z value by c, the speed of light. The Hubble constant (H_0) is a value that keeps getting refined by more precise data from the observational tools cosmologists and astronomers use today. Recent values of the Hubble constant are in the range of about 67 to about 72 kilometers per second per megaparsec [(km/s)/Mpc]. So, with our redshift-calculated speed and the Hubble constant value, we can calculate distance using Hubble's law. This method is useful for determining distances beyond about 100 megaparsecs (Mpc).

D. so, what's in our universe

Using spectroscopy and distance-measuring techniques has led humans to some understanding of what is in our universe. Some things we can observe because we see the visible electromagnetic radiation they emit or reflect. Others we can observe because we have built equipment that can detect their electromagnetic radiation that is outside the visible range. Other things we cannot (at least not yet) observe because they do not emit or reflect detectable electromagnetic radiation or matter particles; these invisible things, however, influence their environments in ways that we can observe.

Things we can see include Earth, its moon and sun, planets and other objects in our solar system and in other solar systems, other stars, and other galaxies.

Some things we cannot see directly, but which seem to exist because of their effects that we have observed, include black holes, dark matter, and dark energy.

Here are some questions about these parts of our universe. Some you may know; others we will answer in class over the next few days; the remainder I leave for you to discover answers (to begin, you might type the question into a suitable internet search program or app and pick appropriate websites from the search results, and then pursue other interesting questions that arise!).

1. Earth

How big is Earth?

How fast does Earth rotate?

How fast does Earth revolve around the sun?

How old is Earth?

What is Earth made of?

What is Earth's magnetic north pole; what is Earth's geographic north pole; are they different?

Does Earth's magnetic field move?

What is precession of Earth's axis?

What causes the seasons on Earth?

What is the imaginary line that divides Earth into northern and southern hemispheres?

2. moon

How big is the moon?

How far from Earth is the moon?

What is the moon's atmosphere?

What about the moon's gravity?

What effect does the moon's gravity have on Earth?

What are the phases of the moon? What causes the phases of the moon?

What is a lunar eclipse? When does a lunar eclipse occur?

Why do we see only one side of the moon from Earth?

3. sun

How big is the sun?

What kind of a star is our sun?

How old is the sun?

What happens inside the sun?

How does the sun affect Earth?

What is a solar eclipse? When does a solar eclipse occur?

What will happen to our sun? What will happen to Earth as a result?

4. other planets and our solar system

What are the planets in our solar system?

How big are the planets?

How far are the planets from the sun?

Why is Pluto no longer a planet?

What else is in our solar system besides the sun and the planets?

How big is our solar system?

Where is our solar system in our galaxy?

How does our solar system move within our galaxy?

What is an exoplanet?

5. stars

What is a star?

Are all stars the same?

What are the life cycles of stars?

What are constellations?

Are all the stars in a constellation the same distance from Earth?

How do we use constellations today?

    6. galaxies

What is the name of our galaxy?

How big is our galaxy?

How old is our galaxy?

Where is our galaxy in our universe?

What are the bulge, disk, and halo of our galaxy?

What is at the center of our galaxy?

    7. black holes

What is a black hole?

Why do we think black holes exist?

What is the event horizon of a black hole?

What is the singularity of a black hole?

    8. dark matter

What is dark matter?

Why do we think dark matter exists?

    9. dark energy

What is dark energy?

Why do we think dark energy exists?

WHAT ELSE DO YOU WANT TO DISCOVER ABOUT OUR UNIVERSE AND US?